D0630460

Oil Producers
and Consumers:
Conflict or Cooperation

Synthesis of an International Seminar
at the Center for Mediterranean Studies
Rome, June 24 to June 28, 1974

Sir Denis Wright, *Chairman*
Miss Elizabeth Monroe and Mr. Robert Mabro, *Rapporteurs*

AMERICAN UNIVERSITIES FIELD STAFF
535 Fifth Avenue, New York, N.Y. 10017, U.S.A.

Library of Congress Cataloging in Publication Data

Oil producers and consumers.

    (Center for Mediterranean Studies conference
series ; v. 1)
    Bibliography: p.
    1. Petroleum industry and trade--Congresses.
2. Petroleum products--Prices--Congresses. 3. Energy
policy--Congresses. I. Monroe, Elizabeth. II. Mabro,
Robert. III. Series: Center for Mediterranean
Studies conference series ; v. 1.
HD9560.1.05   338.2'7'282   74-21726

Library of Congress Catalog Card No. 74-21726
ISBN - 0-910116-87-3

II

Authors' Note

The publication of this report of five days' discussions in Rome has been expedited because of the pressing worldwide impact of the topics discussed. For this reason, and with the prior agreement of the seminar participants, these have not reviewed the manuscript. The authors therefore take responsibility for their synthesis of the discussion, which attributes no particular view to any single participant.

Elizabeth Monroe
Robert Mabro

Oxford
September 1, 1974

Introduction to the Series

Created in 1968 and located in Rome, the Center for Mediterranean Studies has added new dimensions to the work of its parent organization, the American Universities Field Staff. The Center serves as an "academic consulate" for the faculty and students of the Field Staff's member universities, as the headquarters for wide-ranging programs of instruction, and as the site for specially funded seminars on contemporary issues. Under the direction of E.A. Bayne from its beginnings, the Center has established for itself a broad constituency of friends and admirers who have participated in its various programs, one of which is a series of seminars on the social and political foundations of trans-Mediterranean relationships.

The seminar of which this book is a result is one of that series. As Sir Denis Wright observes in the preface, an earlier seminar on a closely related topic also resulted in a book (*The Changing Balance of Power in the Persian Gulf* by Elizabeth Monroe). Thanks to the fine reception accorded that work and other publications, the Center is offering the present volume as the first of a *Mediterranean Series* to be published in paperback. The next volume on problems of Mediterranean kinship and modernization is planned for the spring of 1975.

Alan W. Horton
Executive Director
American Universities Field Staff

Contents

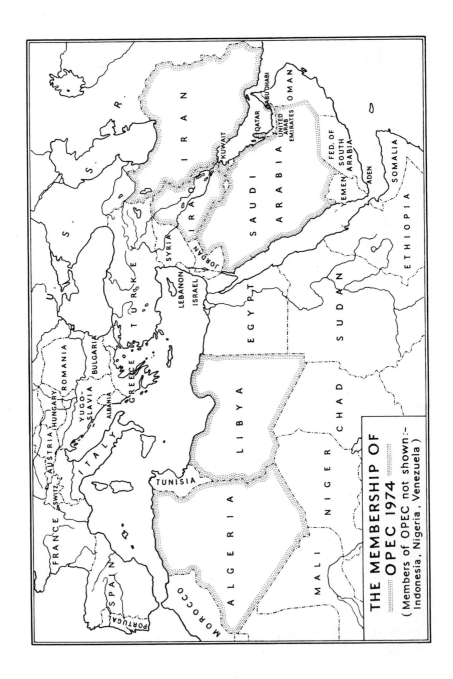

THE MEMBERSHIP OF
OPEC 1974
(Members of OPEC not shown:-
Indonesia, Nigeria, Venezuela)

Preface

The seminar of which the report follows was held in Rome from June 24 to June 28, 1974, under the auspices of the American Universities Field Staff. It was in many ways a sequel to an earlier seminar held under the same auspices in 1972 on "The Changing Balance of Power in the Persian Gulf." Then, discussion of the political vacuum caused by the withdrawal of Britain's military forces from the Gulf in 1971 highlighted the growing dependence of the Western world and Japan, and not least the United States, on Middle Eastern oil. We imagined that these problems, and particularly the currency and investment problems entailed by the mounting revenues of the oil-producing states, would not overburden the world's economic and monetary systems.

At the time, the world had hardly begun to pay heed to these problems, let alone consider how they might be faced. But Mr. E.A. Bayne, Director of the Field Staff's Center for Mediterranean Studies in Rome, proposed early in 1973 organizing a second seminar that would cover both the problems of producer wealth and its investment, and the capacity of the consumer world to handle the huge sums involved. In consultation with Miss Elizabeth Monroe, Mr. Robert Mabro, and myself, an agenda was drawn up for a seminar which we called "Petroleum Producers and Consumers: Towards Mutuality of Purpose." Long before it was due to take place in June 1974, however, the action of the Organisation of Petroleum Exporting Countries (OPEC) in quadrupling the price of oil to the consumer countries compounded problems that were already formidable and gave fresh point to our proposed seminar.

3

The list of participants on p. 5 shows that they represented all the principal actors on the scene: producers capable of spending quickly on their own development; producers with less openings for home investment and an urge to conserve their oil; large consumers from rich industrial countries; smaller, poorer consumers from less developed countries; the major international oil companies; academic economists. I am grateful to them all for finding the time to attend a meeting of which I was proud to take the chair.

My only regret is that representatives invited from Saudi Arabia, Algeria, and Japan were unable to join us. All three states, for very different reasons, are key countries as far as oil is concerned. So is the U.S.S.R., which could be an important actor on the world oil scene; but since big deals with Russia are unlikely in the years 1974-1980 which were our principal concern, we confined this particular seminar to the problems of the Western world and Japan.

In the course of our seminar, many shades of opinion were expressed. This record attempts to synthesize them. It represents the views of no one participant. Miss Elizabeth Monroe and Mr. Robert Mabro, in their admirably concise and lucid report, have done their best to reflect the whole range of opinions advanced during our discussions. It is my hope that at least some of the ideas set out in the pages that follow will provoke further thought and discussion by those seeking solutions for the formidable and complex issues now facing an oil-hungry world. If so our seminar will, as its sponsors had hoped, have been more than just a stimulating intellectual exercise for those who took part in it.

Finally, I should like to offer a sincere word of thanks to Mr. Bayne and his staff at the Center for Mediterranean Studies for the excellent arrangements made for us during our week-long stay in Rome.

Denis Wright
Haddenham, Bucks   September 1974

Seminar Participants and Observers

Chairman: Sir DENIS WRIGHT (Great Britain), diplomatist. Former Ambassador to Iran and Ethiopia; Director, The Standard and Chartered Bank, Shell Transport and Trading Company, The Mitchell Cotts Group.

Rapporteuse: Miss ELIZABETH MONROE (Great Britain), historian. Fellow Emeritus of St. Antony's College, Oxford; staff of *The Economist,* London. Author of *The Mediterranean in Politics, Britain's Moment in the Middle East, The Changing Balance of Power in the Persian Gulf, Philby of Arabia,* etc.

Rapporteur: Mr. ROBERT MABRO (Greece), Fellow and Senior Research Officer in Middle East economics, St. Antony's College, Oxford. Author of *The Egyptian Economy, 1952-72,* etc.

———

Mr. MANUCHEHR AGAH (Iran), economist. Under Secretary, Ministry of Finance, Imperial Government of Iran; former chairman, Faculty of Economics, University of Tehran.

Mr. E. A. BAYNE (United States), political journalist. Director, Center for Mediterranean Studies. Author of *Four Ways of Politics, Persian Kingship in Transition,* etc.

Mr. ROLAND DE MONTAIGU (France), industrialist and engineer. Staff of the President, Compagnie Française des Pétroles; former Director, Middle East Department, Compagnie Française des Pétroles.

Mr. KHODADAD FARMANFARMAIAN (Iran), economist. Chairman, Bank Sanaye Iran; former Governor, Central Bank of Iran and Managing Director, The Plan Organization.

Mr. JOHN H. FORSYTH (Great Britain), economist. Economic advisor, Morgan Grenfell and Co., Ltd., London.

Professor ROSE L. GREAVES (United States), historian. Department of History, University of Kansas; former staff historian, British Petroleum, Ltd.

Professor ARMIN GUTOWSKI (Federal Republic of Germany), economist. Member, Council of Economic Advisors to the Federal Government; chief economic advisor, Kreditanstalt für Wiederaufbau; Institute of Economic Sciences, University of Frankfurt.

The Honorable ABDLATIF YOUSEF AL-HAMAD (Kuwait), Director-General, Kuwait Development Fund.

Dr. R. M. HONAVAR (India), economist. Minister for Economic Affairs, Indian High Commission, London.

Mr. WALTER LEVY (United States), petroleum consultant. Former advisor to the President's Materials Policy Commission; member, National Security Resources Board and Policy Planning Staff.

Mr. G. W. MACKWORTH-YOUNG (Great Britain), banker. Vice Chairman, Morgan Grenfell and Co., Ltd., London.

Mr. PETER H. R. MARSHALL (Great Britain), diplomatist. Assistant Under Secretary of State for Economic Affairs, Foreign and Commonwealth Office.

Professor JON B. MCLIN (United States), political scientist. Senior Associate for Western Europe and International Organizations, American Universities Field Staff. Author of *Canada's Changing Defense Policy, 1957-1963*, etc.

Professor LJUBISA PARADJANIN (Yugoslavia), engineer. Professor of Engineering, University of Belgrade; Chairman, Yugoslav National Council for the World Energy Conference.

Mr. CARMICHAEL CHARLES POCOCK (Great Britain), industrialist. Managing Director, Royal Dutch-Shell Group of Companies.

Mr. MICHAEL POSNER (Great Britain), economist. Fellow and Director of Studies, Pembroke College, Cambridge; formerly economic consultant to H.M.'s Treasury and to the International Monetary Fund. Author of *Economics of Fuel Policy, Italian Public Enterprise* (with S. J. Woolf), etc.

Mr. STEPHEN POTTER (Great Britain), economist. Director, Balance-of-Payments Division, Organization for Economic Cooperation and Development.

Mr. RENÉ SERVOISE (France), economist. Minister Counselor of the Embassy of the Republic of France in Rome; former Ambassador of France to the Democratic Republic of Vietnam.

Dott. STEFANO SILVESTRI (Italy), political scientist. Deputy Director, Istituto degli Affari Internazionali and author of *La Sicurezza Europea, La Strategia Sovietica, Il Mediterraneo,* etc.

Ambassador PHILLIPS TALBOT (United States), political scientist. President, The Asia Society; former envoy to Greece and Assistant Secretary of State for Near Eastern and South Asian Affairs.

Mr. RICHARD HENRY TURNER (Great Britain), banker. Adviser on Middle Eastern Affairs, Bank of England.

Dr. P. VAN DEN BERG (Netherlands), attorney. Senior petroleum consultant, Organization for Economic Cooperation and Development; former Secretary, Iranian Oil Operating Companies.

Professor RAYMOND VERNON (United States), economist. Herbert F. Johnson Professor of International Business Management, Graduate School of Business, Harvard University; director, Harvard Multinational Enterprise Project. Author of *The Economic and Political Consequences of Multinational Enterprise, Sovereignty at Bay,* etc.

Professor JOHN WATERBURY (United States), political scientist. Associate for the Islamic Mediterranean, American Universities Field Staff. Author of *Commander of the Faithful, North for the Trade,* etc.

Mr. MINOS A. ZOMBANAKIS (Greece), banker. Vice Chairman, First Boston Corporation, and Managing Director, First Boston (Europe) Ltd.; formerly Senior Vice President, Manufacturers Hanover Trust Company.

Introduction
Causes for Worry 1974-1980

Oil is a commodity like no other. It is a prime source of energy that cannot be quickly replaced. Its production is held in few hands. Its producers can choose at the turn of a valve either to increase its flow or to conserve it in the ground.

On October 16, 1973, a world that was consuming oil with abandon got a sudden shock. The Organisation of Petroleum Exporting Countries (OPEC) announced that it was doubling the price. For over two years previously it had been edging this price upwards by negotiation with the international oil companies. It had been prompted to do so for a number of reasons, among them two devaluations of the dollar and mounting inflation of the cost of its members' imports. But the unilateral jump was unexpected, and the shock that it caused was the greater because it happened to coincide with the Arab-Israeli war. Most, though not all of OPEC's members are Arabs.[1] These Arab members (OAPEC)[2] played their part in the war by imposing a boycott, directed chiefly at the United States, but extending to all of Israel's friends. Panic buying followed and led, on December 23, to a further doubling of the price, announced by the Shah of Iran. The era of cheap oil was over, and as from January 1, 1974, oil was costing its consumers four times more than it had done in the previous September.

The dimensions of the economic and financial change caused to the world have as yet scarcely penetrated the public mind. Oil producers are taking some time to realize the extent to which they have become the new rich, and the big industrial consumers of the Western world and Japan (who are in varying degrees the new poor, depending on whether or not they produce any oil at home) seemed, once the immediate emergency of the Arab embargo was over, to be ready to resume many old consumption habits, and to view their plight with undue complacency. Their view will or ought to change as the oil bills for 1974 come in.

The amount of the producers' wealth is a guess depending on price and pace of production. What is certain is that the additional imports that they can absorb immediately will account only for a small proportion of their new earnings. At minimum OPEC as a whole will, after satisfying its import needs, dispose of a surplus of 65-70 billion* dollars for the single year 1974. Cumulatively and including interest, the sum could be $650 billion by 1980 and $1,200 billion by 1985.[3] To give an idea of scale, total United States overseas investment today is $90 billion.

Consumers long took cheap oil for granted. They were in no hurry to exploit or invent alternative sources of energy. None of these, except coal for power plants, can make any difference to the role of oil before 1980, if then. For at least five years, OPEC has things its own way, is able to earn more than it can spend, and, unless its members are introduced to ways of disposing of their funds safely and profitably, may prefer oil in the ground to dollars in the bank.

Consumers are therefore confronted not only with high prices, but with the possibility of oil shortages brought on by conservation. They are of two minds about what to do to keep the oil flowing. Some say that the new price cannot be met without economic chaos, and that it must be brought down by adopting crash austerity programs while making all haste to develop known sources of oil outside OPEC — in Alaska for the United States, in the North Sea for Europe, in Eastern Siberia for Japan. Others believe that

*The word billion is used throughout this report to mean 1,000 million.

the high price, together with the collapse of the unprecedented economic boom of 1972-73, will between them so much curtail demand that consumers need do nothing, as the price will fall anyway. Yet others, and they seem to be the bulk of informed consumer opinion, commend austerity and a stop to waste, but reckon that a serious shortage of oil would be so disastrous that they must devise means of keeping production going, inevitably at prices much higher than those of September 1973. Their aim is to meet the producers by laying before them satisfactory ways of investing their surplus funds.

Even worse hit than the industrial consumers of the West and Japan are the less developed countries of Asia, Africa, and Latin America. Though small consumers, their populations — often huge, rapidly multiplying, and sprawling — live close to the breadline, with no latitude for austerity. They cannot afford growth at the new price, and no oil for their agriculture and transport today means no food for tomorrow. They need rescue rather than aid.

This crisis strikes all consumers at an unfortunate moment. The whole world is beset by inflation, and many who are not major exporters of oil or of some other highly priced raw material are having difficulty in balancing international payments. (Britain's deficit on current account, for instance, is larger than it has ever been before; Italy is in even greater difficulty.) Paying for oil at the new price will aggravate their plight, and a recession, if as imminent as many believe it to be, would cause them new difficulties. In theory, governments of the Western industrial world and Japan could relieve these stresses by imposing austerity on their people, or by otherwise manipulating their home economies by means of budgetary and tax devices. In practice, most have weak majorities and are too fearful of unemployment to feel they can ignore their electorates. They are therefore loath to explain to these the need for the significant changes of habit, and in their industrial and banking structure, that will be necessary in order to pay the oil bill for the 1970s. Japan seems to be the first of these

11

nations to publicize the need for discipline on the part of its workforce.[4] The measures finally adopted may be as rigorous as those imposed at the end of the Second World War.

On the face of it, the relationship between oil-producing and oil-consuming states has become one of conflict. This report will argue that there is more common ground, and more likelihood of cooperation between the two, than would at first appear.

[1]The non-Arab members are Indonesia, Iran, Nigeria, and Venezuela.
[2]OAPEC consists of OPEC's Arab members, plus Bahrain, Dubai, Egypt, and Syria.
[3]International Bank for Reconstruction and Development (IBRD) estimates in a report on increased funds for underdeveloped countries, July 1974; quoted in the London *Times,* July 31, 1974.
[4]See Saburo Okita, "Natural Resource Dependency and Japanese Foreign Policy," *Foreign Affairs* (New York) July 1974, pp. 714-724.

Supply, Demand and Price

There are three actors on the oil scene of 1973-74. They are the producer governments, the major oil companies, and the governments of the consumer states.

The first belong to OPEC, and operate there in a unity hitherto unimpaired by their very different characteristics and domestic needs. Some of them have large populations, resources other than oil that they are eager to develop, and standards of health and education that have long since started to rise. Indonesia, Nigeria, Iran, Algeria, and Venezuela, respectively with populations of 127 million, 65 million, 30 million, 14 million and 10 million, intend to diversify their economies and turn themselves as soon as possible into industrial centers that will before long vie with the West and Japan. Iraq, also with 10 million inhabitants and with promising alternative resources, is entering the same category and settling down to consistent planning and implementation of its plans. These states look ahead to short horizons and fast targets.

Another group is differently placed. Its members have sparse or small populations and little or no alternative development base. They are the largely desert states of Saudi Arabia and Libya and the small Emirates on the Arabian shore of the Gulf, — Kuwait, Qatar, Abu Dhabi, Dubai. These, though they have short term ambitions in the field of home education and welfare and intend to establish such industries as they can render viable, have in the main a much longer horizon. Their governments aim to maintain the living standards of the oil age for future

13

generations by making investments abroad; they want to buy continued well-being for these and at the same time to acquire assets in friendly foreign states that might contribute to their own political survival.

This brief categorization, amplified in Chapter III, highlights the producer governments' major problem: all still lack the social infrastructure immediately to adjust to their new level of income, but some can adjust sooner than others. No producer can at once spend all that he earns at the new price of oil; some will want quickly, others more slowly, the assets that all intend to buy. All will have, during an interim period that varies in length, either to lend the balances that they cannot spend, or else to conserve their oil in the ground. It so happens that by far the biggest surplus funds are going to accrue to the state that can spend them least quickly — Saudi Arabia.

The second actor is the major oil companies. For years these great national and multinational concerns — American, British, Anglo-Dutch, and French — virtually managed the world oil industry. They financed the risks of prospection, estimated the end-demand by consumers in order to determine the rate of production, judged the competition in the market that set the price of products, delivered them all over the world, and sold them at the pump. They decided where production should be high or low on a basis of where it was most economic, and shared out available supplies when oil was short. After the Second World War, a number of small independent companies intruded on their provinces in the Middle East, somewhat reducing their domination of its major industry.

Oil was in plentiful supply in the 1960s. In the course of 1970-71 a tight supply situation led to a radical change in OPEC's conception of future supply and demand. World demand for its oil (see table 1), particularly from the United States, was rising at an astronomic rate and, at the hands of the companies, this demand was being met from only a handful of sources. The companies were not discovering new reserves fast enough to keep pace with Western

TABLE 1

	Oil Imports 1970-73 (thousand barrels daily)			Rise: 1973 over 1968 (per cent)
	1971	1972	1973	
U.S.A.	3,930	4,740	6,205	17.2%
W. Europe	13,520	14,065	15,310	7.9%
Japan	4,720	4,815	5,760	13.5%

(Source: B.P. Statistical Review of the World Oil Industry, 1973.)

---

and Japanese appetite. The ratio of Middle East production to reserves was increasing. The producers saw that their asset was becoming yearly more precious, and understandably sought to raise its price. The companies concurred, and in a series of negotiations beginning at Teheran in January 1971, and continued throughout 1972-73 at Tripoli, Baghdad, and Geneva, gradually raised the price by negotiation, sometimes absolutely, sometimes to allow for successive devaluations of the dollar.

During the early 1970s, some extraneous events further changed the producers' perceptions. The Club of Rome proclaimed that Doomsday was in sight, and that the world would run out of certain raw materials by the end of the century. In America, experts began to warn an unheeding public that dependence on the Middle East for oil was already at a "peril point" envisaged for 1980.[1] Decisions in America and Europe to revalue or float various currencies worried OPEC members as to the value in real terms of payments for their oil. In the course of 1972, two of those least in need of immediate funds — Kuwait and Libya — decided to practice conservation.

Also in 1972, the major oil companies were tackled by OPEC on another count. Its members sought participation in the ownership of the concessions. Again the companies reluctantly concurred. Their agreements with the different producers were not identical. As a start, many of these agreed to a 25 per cent participation arrangement devised by Saudi Arabia and accepted in principle by most of the Gulf states.[2] Iran — odd man out in this case — made its own arrangement. Its National Iranian Oil Company is much the most experienced of the local companies; NIOC took over all operations in the international companies' area, while they, in return for some 20-year buying privileges, agreed to act merely as a production contractor for NIOC. Again price was affected. Governments, on acquiring a proportion of their own oil to sell, became aware that they could get more for it on the open market than the companies were paying them in royalty and tax, or for the "buy back" oil that the companies purchased from them. Finally, impatient with long haggling over price, OPEC took a plunge, and without consulting the companies, on October 16, 1973, meeting in Kuwait, doubled the posted price of oil as already described.[3]

Its members agreed to charge a price that they justified by the fact that the market price had risen in relation to posted prices and that adjustment was necessary. The posted price of oil had long borne no relation to its cost of production; there was no criterion whereby to judge how much adjustment to make, and OPEC set a price that it thought consumers would tolerate. It was more than right. The Arab-Israeli War was in progress and, to further the Arab cause, its Arab members (except Iraq) imposed a boycott in the form of a 10 per cent cutback in output, scheduled to be further cut back at 5 per cent per month till they achieved certain political targets connected with Palestine; this boycott they later modified in respect of deliveries to their friends, but made total in the case of two countries marked down as friends of Israel — Holland and the United States.[4]

During the ensuing panic about short-term supplies, consumers paid on the open market as much as $17-18 per barrel, and on one occasion $20, for oil that had before the

16

price rise and the embargo cost an average $3 per barrel. Noting the prices that people were prepared to pay to get oil, the Shah of Iran took another initiative. He convinced most of his OPEC colleagues that it would be safe to double the price again; they did so as from January 1, 1974.

Again the companies had to concur. In theory a high price suits a commercial firm engaged in expensive operations all over the world provided that its profit margin remains; but though in fact the companies' net incomes rose enormously thanks to the 1974 prices, they were worried about the reduced demand these prices presaged, and about the changing shape of their industry. A few individuals in the oil-producing states, notably in Saudi Arabia, were also worried; some Saudis feared that too high a price would encourage substitution and impair their long-term interest in selling oil from their immense reserves; they also judged that it was unwise to contribute to economic recession in the industrial countries and to jeopardize economies on which the oil producers depend for expertise and industrial supplies. (A latent additional rise in the producers' earnings took place when the participation proportion rose to 60 per cent, since the bulk of the oil was thereafter being sold on the basis most advantageous to them.) When the panic about supply abated, market prices fell from their winter peak and, as this report is being written, are fluctuating at around $10 to $11 per barrel; $10 is the price at which, for convenience, the calculations in its pages will be made. At this price, there were clear indications of a fall in steady demand by mid-1974 (see Appendix, Table III). Customers paid it in order to restock after the embargo. Supplies were maintained beyond the crisis period, so that by the summer of 1974 tankers were loading about three million barrels of oil every day more than was being consumed at the other end of the journey, and it was becoming difficult to find anywhere to place these surplus stocks.[5]

The natural ceiling to oil's price is the cost of substitutes, but substitutes, unfortunately for consumers, are simply not available, and cannot quickly be made so. America's Federal Energy Authority hopes to make "Project Independence" operative by 1980, and the British are banking on self-sufficiency from the North Sea by about that date, but other states are not so well endowed.

Substitute energy is easier to come by for power plants than for transportation, but with the single exception of *coal* used as coal in existing installations, substitution cannot be brought about immediately. Plants take time to alter from oil- to coal-burning, and coal supplies (see table 2), though a cheap alternative in the United States, raise some familiar manpower problems. It is calculated that world coal trade in 1973 (some 100 million tons) could be raised only to some 200 million tons by 1980. But it is also expected that substitution will not make much difference to the price of oil until three or four choices of substitute are on the market.

TABLE 2

World Coal Reserves

Soviet bloc	61½%
North America	17%
Asia (mainly China)	17%
Europe	2½%
	(of which Britain .08%)
Africa	1½%

Australia and Latin America each less than ½%.
(Source: Hill and Vielvoye, p. 182)

Of these choices, that of which most is expected is *nuclear energy*. Its development is on the way, but has lagged owing to the high capital cost of the vast apparatus needed for its generation, and owing to the cheapness of oil. So far, development is advanced only in a few wealthy states.

Expansion has also been held up by pressure from environmental enthusiasts who obstructed the search for sites at home, and by government reluctance to spread abroad techniques that can be used for the manufacture of nuclear armaments. The most positive steps toward using nuclear fission in atomic power stations have been taken by the Americans with the French, Germans, and Japanese a considerable way behind; Britain advanced more slowly out of indecision until mid-1974 as to what type of reactor to adopt for the next generation of nuclear power stations; Italy still has no proper nuclear plan. A handicap which began to loom in 1974 was a potential shortage of enriched uranium, of which by far the largest supplier is the United States.[6] *Nuclear fusion* — a process whereby fusile material must be heated to over 50 million degrees centigrade — is so far only at the experimental stage and will probably make no impact before the twenty-first century.

Natural gas tends to be available in the same areas as oil. The sight of it flaring to waste in deserts is familiar to anyone who knows a remote oilfield. Where it occurs in consumption areas, it is cheap, provided that it can be easily piped. The United States uses it extensively already; an EEC set of revised energy targets published in mid-1974 proposed an increase in the proportionate use of gas from 2 per cent in that year to 25 per cent in 1985. When natural gas is extracted far from consumption areas, it needs liquefaction, and liquefied gas is extremely expensive to carry.

Coal gasification, though unlikely to materialize before 1985, is expected to cost the equivalent of oil at $12 per barrel in terms of 1974 dollars. The use of coal for producing synthetic oil would cost more — say $15 per barrel in Europe or Japan, though less in the United States.

Tar Sands, of which Canada owns immense quantities in Athabasca, call for vast processing equipment and use of energy. Venezuela, too, owns them in substantial amounts.

Shale oil, available in quantity in three states of the United States (Colorado, Utah, and Wyoming) and also in

Brazil and China, entails the use of huge quantities of water and the waste disposal of a massive residue of friable dust that causes environmentalists to oppose its processing absolutely.

Local developments such as the harnessing of *solar energy, tides* or *wind,* and tapping the *geothermal heat* from underground are in operation here and there. America has gone some way towards devising an apparatus that uses solar power for domestic heating; Reykjavik warms itself by geothermal energy; France has constructed in Brittany a barrage powered by tidal surge; windpower, though intermittent and unreliable in agreeable spots, could be used much more widely than at present.

Some processes, though expensive, may be justified in certain countries on grounds of self-sufficiency for strategic reasons. A snag to all expensive programs is that the private sector is reluctant to invest in them because they could overnight be rendered uneconomic by a reduction in the price of oil. Only governments can sponsor them on a scale large enough to mitigate the oil bill, or to strengthen their power of bargaining with the producers.

A more immediate ceiling to price is the producer's grasp of the risk he runs by setting it too high. This risk is not merely the encouragement of search for substitutes and for finds in new areas; OPEC producers with relatively small reserves may welcome and subscribe to these, as they would prefer to distribute more widely than at present the responsibility for supplying the world with energy. Some see a greater risk to themselves in the impact of high prices on strong economies, and do not want to jeopardize this strength and run the risk of creating worldwide economic havoc.

OPEC's price fixing problem is not easy. The short-term price that it fixes cannot be divorced from its estimation of future demand, and this must be a guess. For instance, a serious recession in the rest of the world could substantially reduce consumption. OPEC might then see oil in the ground as less valuable, and bring short term prices down in order to try and generate Western recovery.

Again, disparity between producer targets and interests affects long-term judgments about optimum price. If one of the more ambitious and overpopulated producing states were to wish to develop fast, and therefore to double its oil output without reducing price, its project would work only if producers less pressed for time and money were to agree to stand down, and preserve the price level by confining themselves to making up the short-fall. Maybe desert producers would welcome this chance to conserve their oil; but the sight of a shrunken share of the market might be too much for their people, and render them uneasy about conservation; disagreement within OPEC about market shares could cause its unity on the score of price to crack.

Again, the sudden appearance of a really big new find somewhere in the rest of the world might crack OPEC solidarity. But at present this is holding, and the only pre-diction that can be made with certainty is that, if market prices fall, the host government's tax-take will be the last element to be reduced. At a guess, the Mediterranean producers and Nigeria, where governments are selling much of their oil directly, might be the first to act indepen-dently. But speculation on such differences is premature.

The oil companies are no better able than the consum-ers to influence price unless they discover substantial new resources outside the OPEC areas. They are looking for these, mainly in "safe" political areas, and are at present spending $4 billion annually around the world on explora-tion, without counting North America. They are also pros-pecting in "risky" areas such as offshore Vietnam and offshore West Africa, where as yet there are no accepted systems for international financing or political guarantees.

In their old fields of action within the OPEC area, the oil companies are in a position to negotiate only on such topics as whether the "commercial terms" of payment asked shall entail sight payment, or on what terms, if they pay the full cost of prospection on behalf of national oil companies, and thereafter hand over exploitation, they shall purchase oil from those companies at discount prices. For the present, most of the oil transportation, refining, and marketing business remains in their hands, but producers have made it clear that they mean to buy their way into these activities, or even to take the companies' place. Faced with these complications, the companies are proving adaptable. In times of great uncertainty, they prefer to keep a low profile in the OPEC states. Several have made provision for the future by diversifying into other energy fields. Shell and Exxon, for instance, have entered the nuclear and coal businesses; Mobil has diversified more drastically by buying into the American mail order business of Montgomery Ward.

All told, the likelihood on the price front is that in the short run till 1980, prices will remain constant or fall very slightly. In money terms, they are likely to rise if inflation continues as predicted. They will continue to be dictated by OPEC, which so far feels successful, and they will bear no relation to supply since OPEC can control this by practicing the conservation policy described in the next chapter, raising the price at will for a commodity in short supply, and thereby maintaining a consistent income.

[1]See, for example, James E. Akins "The Oil Crisis: This Time the Wolf is Here," *Foreign Affairs,* April 1973.
[2]The percentage was by subsequent negotiation raised to 60 per cent for most producers.
[3]The "posted price" was an artificial price till then fixed by the oil companies, and was the basis on which the tax they paid to the host governments was calculated.
[4]The embargo on the United States was raised in March 1974, in recognition of Dr. Henry Kissinger's peace-making efforts between Israel and the Arab states; that on Holland not until July.
[5]To predict how long this situation will last or whether it will affect price is impossible; it could be seasonal, and end in the course of a hard winter.
[6]A note by the U.S. Atomic Energy Commission published in the *Federal Public Record* (August 1974) states that it is unable to accept any new contracts for supply.

II.

Conservation and Austerity[1]

Conservation by the producers got much publicity when it took the form of an Arab oil embargo, but conservation is not necessarily a political sanction. Some producers favor it — indeed, have begun to practice it — for domestic reasons of their own. Kuwait and Libya did so from 1972 because their known reserves are limited and because, having small populations and limited capacity for greater domestic development, they could afford to restrict production even before the price rise of 1973 increased their wealth. The ratio of reserves to production is also low in Algeria, Venezuela, and Indonesia, and may be so in Iran, but all these states have large populations hungry for industrial and social development, and have to set judgment about conservation against popular expectations calling for heavy expenditure as soon as possible.

Where the motive for amassing funds is scanty, conservation is an obvious course. Producers no longer see reason in stepping up production simply to sate appetite for oil in the consumer countries, and to hoard the money proceeds in banks, largely for transfer to other banks. They have come into these huge surplus funds only in 1974, and need time to sort out their ideas about how best to protect future purchasing power.

Most of them are thinking hard, but many are pulled two ways, and beset by hestitation. All the desert producers are Arab, and Arabs have their good name with the poorer Arab states to consider — notably with Egypt. The King of Saudi Arabia wants to preserve his royal image as the leader of the Arab and Islamic worlds, and to match Iran as a power with influence in the Gulf. Generous lending and massive armaments are expenditures that match these aims. Yet conservation might serve him well on other counts. The dilemmas that face him are described in Chapter III.

On the other hand the desert producers, though run by élites less affected by the groundswell of public opinion than are the Western consumers, have home political acceptance to consider, and the instincts of some of their publics are for conservation; public opinion as represented by the elected Assembly in Kuwait thinks it sensible to produce as little oil as is consistent with national security and independence. All producer states are training armed forces which, through education in handling sophisticated weapons, have acquired minds of their own and may want a say in oil policy. In a word, the level of production in all OPEC states has nothing to do with cost, and has become a purely political judgment taken by people whose political aims and requirements are not identical. OPEC governments can decide whether and when to produce or conserve, and may or may not do so unanimously. Owing to the inevitable effect of shortage on price, those who conserve need not lose financially.

To all political judgments there is a political reaction. Were the producers to overdo conservation (say, to impose a prolonged embargo mounting at 5 per cent per month up to a level of 40 or 50 per cent) or were they compelled by regional war or domestic rebellion to cut production in a big way, the consumers — judging by some casual American talk — might contemplate military action to restore output. Such an operation would not be impossible to mount.

Yet recourse to force would be unwise. A major point that the invading consumer would have to consider is the likelihood that invasion would cause fellow members of OPEC to rally round the victim of his aggression. Quite apart from the political effect of the step on superpower relations, it would affect regional opinion and for an appreciable time certainly sever the consumer world from the source of supply attacked, and possibly from all OPEC sources. A producer so assaulted is determined to apply a scorched earth policy. Destruction of apparatus such as jetties, pipelines, and offshore rigs is easy; techniques for the temporary destruction of wells are also advanced, and the injection of cement into bore holes puts wells out of commission for months.

Producers are alive to the risk of overacting. The Arab embargo of 1973 was most cautiously applied. Within a matter of weeks, it was lifted in the case of all "enemies of Israel" — a phrase liberally interpreted to include all Western Europe except the Netherlands;[2] after a pause to secure a declaration of approval of the Arab cause, it was also lifted against Japan.

Whether for political or economic reasons, cutbacks that all producers can (for the years discussed in this paper) operate without domestic hardship are inevitable sometime and somewhere; their size and site is difficult to predict because all producer states are pulled both ways.

Consumer reactions to conservation are happily not all as drastic as resort to force. Some have materialized already, triggered by the Arab embargo as well as by the rise in price. One is the speeding up of production outside OPEC, the bulk of it in major consumer areas.[3] In North America in particular, wells that had been shut down as uneconomic with oil at the old price became worth reopening at the new; also, former ecological and other objections to the Alaska pipeline melted away. Two million barrels per day from that source are expected by 1977. In Europe, a British contribution of an ultimate three million barrels per day from the North Sea is expected to start in a few years' time. Norway, if it were to adopt the same policy as Britain, could contribute to the rescue of Western Europe, but for the present prefers not to disturb the pattern of its domestic economy. Japan, the biggest single importer of crude oil in the world, is hoping to diversify its sources of supply by importing less from the Middle East and more from Eastern Siberia.

Cooperation among consumers could speed the austerity process, particularly over technical research into economy devices. So far, cooperation has been practiced chiefly over sharing supplies in the event of shortage. The development of a scheme for this purpose was recommended at the World Energy Conference of February 1974, and as its outcome an Energy Coordinating Group met in Brussels in July 1974.[4] Consisting of official representatives, this meeting reached "substantial agreement" on sharing, in an emergency, oil from both domestic and imported sources; its findings have, at the time of writing, yet to be adopted by governments.

Another likely consumer reaction to conservation by OPEC is more economic use of oil. Present techniques for converting fossil fuels for use in power stations are said to lose 60 to 65 per cent of energy in the process — a figure that could be greatly bettered by anyone ready to pay the capital cost of expensive adjustments. Again, environmentalists keen to purify the air of cities have encouraged the use of nonlead gasoline, which entails a 5 per cent increase in the consumption of crude oil, and of desulfurized fuels for which the increase is 8 per cent.

A nation bent on austerity could go much further without impairing its economy. Some experts argue that if all consumers were to do so because each came to a decision that the price asked in 1974 was "simply not bearable," they could by acting quickly and resolutely, use austerity as a weapon and, coupling it with pressure to speed up other sources of oil supply, force down the price asked by OPEC.[5]

One estimate is that they could, through austerity, reduce the rate of increase in free world oil consumption from the average of 7.5 per cent per annum in 1968-1972 to an average rate of 2.7 per cent per annum for the years 1972-1980.[6] The estimate from the same source for the slowdown that is likely to happen without special effort is 5.1 per cent.

As will be seen in Chapter III, on the production side the attitude of Saudi Arabia alone can govern the supply position; similarly, on the consumption side the attitude of the United States toward economy and austerity can govern reduction of demand. American consumption is far greater than that of any other single country.[7] Just as the average American automobile consumes far more gasoline per mile than the average European or Japanese machine, so the American tractor uses twice as much fuel as the French one to produce a bushel of wheat, and so on. All over the world, a certain amount of economy in transportation, industry, and domestic use can be achieved through voluntary effort, but austerity on a big scale, or a crash austerity program for the rest of the 1970s, can be brought about only by government action and wide publicity.

Possible techniques are legion; they include, for instance, the recycling of combustible waste. Perhaps those best worth mentioning here are compulsory insulation and the use of fiscal policy to enforce economies.

A recent study on heat conservation by the Massachusetts Institute of Technology reaches the startling conclusion that it is cheaper to conserve heat than to produce it. Insulation is therefore a key to lower domestic and industrial consumption; a similar calculation by Shell suggests that economies in these two fields could be as much as 20 per cent and 10 per cent respectively. It is easier and cheaper to insulate new premises than old, but even in the latter, small capital investments (e.g., on thermostats to regulate both heating and air conditioning) yield a quick payoff; they could make a difference in apartment houses where the landlord who does not pay the fuel bill is less likely than is a private householder voluntarily to install insulation. Estimates of the savings attainable by means of full insulation are:

draught-proofing	7%	temperature	
roof insulation	7%	standardization	2%
wall insulation	3%	double glazing	1%

Glass is a poor insulator, and the fuel crisis may be the death of the picture window.

Another method of procuring industrial and domestic economy is by differential tariff. In the United States, public utilities adjust bulk rates upwards; the more consumed the higher the rate. Britain uses the opposite and less economic pattern; there the user above a given minimum pays a lower unit charge.

Results just as significant can be obtained in the field of transportation. In the United States, the consumption of motor gasoline is 35 per cent of total consumption (1970); the corresponding figure for Western Europe is between 16 and 17 per cent, and for Japan 14.9 per cent. The average performance of cars in miles per gallon in a number of countries tells a story about the cost of gasoline — expensive where the mileage per gallon is high:

France	25 mpg.
Italy	20 mpg.
Germany	15 mpg.
United Kingdom	15 mpg.
USA	8 mpg.

The difference in cost is not that of the oil component but of the tax imposed by the home government, punitive in the case of countries that applied it in order to maintain the demand for home-produced coal. (The peculiar differential so created between the yield of tax to the consumer government and the yield to the producer government in tax and royalties was long a sore point in the Middle East.) Already, under the impact of high oil prices, the American taste for huge cars is changing; it was created by distance to be covered, snobbery, an excellent network of fast highways, and the greater profit earned by the manufacturer on a big car than on a small one. The percentage of small to large car manufacture in the United States three years ago was 35 per cent; the figure is expected to be 60 per cent in 1975. Taxation on gasoline, being indirect, has the social drawback of hitting rich and poor equally; a tax on performance measured by horsepower, cubic centimeter output, engine stroke, and so on, can be levied on a more equitable basis.

Obvious economies are also obtainable by imposing speed limits, by subsidizing public transport, particularly for commuters from suburbs into towns, and by curbing one very popular practice — the use of road tax for the building of fast highways; the United States government blocked the project for carrying interstate highways into Washington. Most of the argument in favor of such economies has been current since long before the price rise,[8] but has fallen on deaf ears because the steps entailed are unpopular. The alacrity with which car-owners in all countries resumed previous habits once the speed limits imposed during the Arab boycott were lifted suggests that much good government publicity, as well as restrictions applied regardless of public opinion, will be needed if austerity is to affect consumption to the extent of bringing the price of crude oil below $10 per barrel.

---

[1] For some forecasts of "normal" and "restricted" supply and demand, see Appendix, Tables II and III.

[2] A main objection was to pro-Israel statements by Netherlands cabinet ministers.

[3] In 1973 North America's share of total world production (including 17.8 per cent produced in the Soviet bloc and China) was 22.8 per cent. (Source: BP Statistical Review for 1973.)

[4] The consumers represented were the United States, Canada, Japan, Norway, and all the members of the EEC except France.

[5] See Walter Levy, "Implications of World Oil Austerity." (Unpublished). February 1974.

[6] Levy. *op cit.*

[7] In 1973, 29.5 per cent of the world total; the Soviet bloc and China together consumed 15.7 per cent; next came Japan consuming 9.7 percent. (Source: BP Statistical Review for 1973.)

[8] e.g., at the conference preceding that of which this book is the record. See *The Changing Balance of Power in The Persian Gulf* (1972), pages 38-9.

III.

The Producers' View

The producers are pleased about their successful moves over price, but anxious about the responsibilities that their wealth entails. They no longer wish to dwell on grievances of the lean years in which they reckoned that they subsidized the West's inordinate appetite. They now have other things to think about.

One of these is their conviction that the industrial consumers can pay the price they are asking. Their image of the West is of competent, experienced peoples having innate ability to mobilize resources. They have watched Western nations get together in emergencies, pay for space programs, summon the means to fight wars in Korea, Algeria, and Vietnam. They acknowledge that paying bills for oil is not so compelling a motive as paying for prestige or for war, but when they scan Western statistics of gross domestic product and see that the sum represented by the new price of oil is only some 3 per cent of that total, they reckon that the proportion is manageable. To any Westerners who argue that the consumers have difficulty in paying the new prices, because to do so their governments will have to dragoon their people and change their industrial structure, the producers answer that they cannot believe Western governments to be so lacking in stability and moral courage as to have lost control over their instruments of employment. They reckon that countries which say that they cannot pay are ill organized, have capacity, talent, and efficiency that they are not using, and are capable of mending their ways.

Some of the producers' convictions about the funds they are earning are equally firm. They see the money as the alternative asset to their major resource — in some states, their only resource. They are parting with a real asset, and want real assets in return, not a mere paper transfer from the credit to the debit side of a ledger. They know that they cannot at once take full delivery in real resources. Neither their skills, their infrastructures, nor their economies are ready to absorb it. They see that they must accept IOUs for some of the goods and services that they are going to need later. Meantime their worries about disposing of surplus funds are not the same for the populous as for the desert states. Differences that were mentioned in the context of conservation are capped by other differences arising out of the time schedules on which OPEC's members plan their respective futures.

Some producers have a short horizon, long-standing ambitions, and fast targets. These are members with large populations, rapidly rising standards of health and education, alternative resources, and intentions to diversify their economies and turn themselves into industrial centers able before too long to vie with the West; such are Indonesia, Nigeria, Iran, Algeria, and Venezuela. Iraq will enter the same category as soon as its ruling few pursue consistent national development. These countries can make and are making immediate purchases, some of armaments on a large scale, all of real assets basic to development — for instance, apparatus for roadbuilding, water storage, trucking, public transport, and agriculture. Iran, for example, boosted investment in its fifth five year development plan (1973-1978) from $36 billion to $68.6 billion in August 1974, and announced that whereas the first year's economic growth rate was to be 11 per cent per annum, a growth rate of nearly 26 per cent (at constant prices) was envisaged for the plan's final year. The announcement shows that for lack of infrastructure, really big orders for industrial products will not materialize for a few years. Until perhaps 1977 or 1978, all producers' financial surpluses will therefore remain vast (see Appendix, Table I), though Indonesia and Algeria have substantial debts to repay and Nigeria has war damage to repair. But by 1980 most will be absorbing, from the main consumers of their oil, goods and services that will mop up a large part of their surplus earnings from that source.

Producers with a long horizon are those with sparse or small populations and little or no alternative development base; they are Saudi Arabia and Libya and the small Emirates on the Arabian shore of the Gulf. Some targets here are also industrial, since industry can be cheaply run where land is easy to come by and natural gas can be supplied free of charge; but for lack of indigenous raw materials and of much water apart from distilled sea water, the projects envisaged almost all have to do with oil — with its refining, transportation, and the manufacture of petrochemicals. One urgent objective of these lands is to maintain the standard of living of the oil age for future generations. They would not forgive themselves if, like the successors of the *conquistadores*, they were to squander sudden wealth in a generation. They want to buy some form of annuity for their grandchildren. A second urgent objective is political survival, and in seeking to purchase this they share some targets with their larger neighbors.

Producers of both kinds live close to countries populated by millions much poorer though potentially more powerful than themselves, and realize that they constitute rich and covetable real estate. A portion of their immediate expenditure is likely to take the form of the Marshall Plan — that is, to be based partly on preserving neighbors from ruin and revolution, partly on creating a market for the goods they intend to manufacture, and partly on philanthropy. They mean to help not only their own home publics but those of their region — Egypt, Pakistan, India — and perhaps in the process to acquire some real assets abroad. In this context the rich Arab producers are rather differently placed from Iran, because they belong, as it were, to a club — the Arab League — whereas Iran is a singleton. But all have plans for regional development and welfare.

In this matter, governments that have been rich for a decade or so are ahead of the rest. Kuwait has been the pioneer because it started almost 20 years ago to earn big money in relation to the size of its population, then only 200,000. The Kuwaitis had a window on the world long before the discovery of their oil; as sailors, they were pearl-fishers and boat-builders and traded all over the Indian Ocean and into Europe. They at once shared their oil revenue with others by opening their schools and hospitals to all comers, and in 1960 set up a Kuwait Fund for Arab Economic Development (KFAED) with a capital of $600 million. By 1973 they had made 38 loans on easy terms to Arab states of all political complexions, granting these on a basis of expert project appraisal, not politics. In 1974, they raised the Fund's capital to $3.3 billion (with authority to borrow up to $12 billion more) and, calling it simply the Kuwait Fund for Economic Development (KFED), opened it to all less developed countries.

Iran has also helped neighbors, notably Pakistan, and with the advent of cash surplus, at once made several other gifts and loans. That to India is described in Chapter VI and has a mutually advantageous basis in that India is to repay in the products of the industries that it sets up with the loan's proceeds, so being spared a hunt for markets elsewhere. The system would seem to be worth extending further, though the burden of repayment, if over a short period, is heavy because budding manufacturers cannot immediately achieve export capacity.

Saudi Arabia, the richest potential lender and the one with the longest future as a big earner, has good reason to join in programs for regional development and welfare. As has been mentioned, it wishes to mark its leadership of the Arab world, and, as guardian of Mecca and Medina and organizer of the Pilgrimage, of the Islamic world also. It intends that one day its wealth will spread knowledge and technical prowess to that world as surely as, over 1,000

34

years ago, its ancestors spread the faith. Its alliance with Egypt, forged over the Palestine question, constitutes a meeting of money on one side with manpower and technical experience on the other; by mid-1974 the Saudi government had made Egypt several large gifts. Yet it is confronted with some perplexing dilemmas as it studies the pace at which to spend its vast means. It is being pressed by poorer Arab states to lend quickly, and it wants to build up its political strength in the Gulf *vis-à-vis* Iran. Yet some of its people favor conservation. Again, its avalanche of funds has started to roll, and for lack of a money market, it has no way of recycling them as the American or Eurodollar markets can do, and so is dependent on Western institutions for handling its surplus; yet its King, for the sake of his political image at home, might find advantage in behaving radically towards the West. Ahead lies another heavy responsibility; as potentially much the largest oil producer, Saudi Arabia could, simply by producing or conserving at an exceptional rate, send the oil price up or down as smaller producers cannot do. It shows no sign of using this leverage; for instance, it favors a rather lower price now than do some of its OPEC colleagues partly so as not to handicap its industrial suppliers, but chiefly because its huge oil reserves will preserve their value for longer if consumers are not in too much hurry to develop substitutes. In 1974, at OPEC's conference held in Quito, it was alone in suggesting a lowering of the price; it then gave in to the majority wish to keep this price high, but, sometime, it might want to use its superior bargaining power.

Decisions about these conflicting requirements are compounded by King Faisal's innate caution. The mass of proposals that the world is churning out for his ears, and the number of callers that throng his anterooms, are enough to bemuse anybody. Saudi Arabia needs more time than most to consider the mass of decisions that fall to so wealthy a state; that is perhaps why it has been slow to subscribe to the all-Arab Fund for Economic and Social Development established in 1971 with headquarters in

Kuwait, and intended to give special encouragement to regional multinational projects on the lines pioneered by KFAED.

Subscription to development through United Nations organizations is another outlet that some producers use and others have under review. Over the years, the IBRD has raised $5 million of bonds in six issues on the Kuwait market. In February 1974, the Shah of Iran pledged to the IBRD and the International Monetary Fund (IMF) more than $1 billion to be divided between the purchase of bonds and a loan to the IMF's "temporary emergency facility" devised in 1974 to assist countries with oil-related deficits on their balances of payments, to be repaid not later than seven years after borrowing.[1] In August 1974, when Venezuela lent the World Bank $500 million, the latter's President, Mr. Robert McNamara, remarked that he hoped loans from oil producing states would in coming years run at about $2½ billion per year.

But in some Arab eyes working through international bodies has snags. For one thing, the rate of interest charged by IBRD to borrowers is high — from August 1974, 8 per cent — whereas the Kuwait Fund tailors its interest rate and funding conditions to the capacity of the borrower, charging between ½ per cent and 4 per cent over periods ranging from 15 to 50 years. For another, bilateral loans can be quickly organized; in Kuwait an application can be processed from start to finish in as little as four months, whereas the ponderous mechanism of international organizations may take more than a year to get under way. Lastly, and this is the cardinal consideration, the agencies set up by UN organizations are to all intents and purposes Western in outlook. The members of their governing bodies have weighted voting rights; there is only one representative of a less developed country on the board of management set up to run the new IBRD fund that is to come into operation on January 1, 1975. The oil producers want more say in the disposal of the funds to which they subscribe.

But whatever the producers do in the way of home development, regional development, and investment in Asia and Africa, their surplus funds will be vast for some years, and, in the case of the desert states, for decades. They must therefore look around the world for secure assets and rates of return on investment. They turn naturally where they have always turned — to the industrial consumer states of the West, where their biggest customers are also the suppliers of the goods and services they most need.

When investing in Asia and Africa, OPEC states have a kindred feeling because they can remember what it was like to be poor; they also feel a sense of control over their investment. When investing in the West, they feel less mastery over the process; when they lend there, they want marketable assets, a real rate of return, and security against currency devaluation and nationalization. They might sacrifice something of the second to get the third.

For a generation, their investors, both public and private, have been buying real estate in Beirut and Cairo. The private sector, though it has dwindled by comparison with their public funds, remains large and continues to grow.[2] The producers would like to spread the practice of buying real property into Europe, America, and Japan. They would like to share in financing and managing other real assets — great concerns such as ICI, General Motors, or Mitsubishi. They would like to become partners in new joint ventures; some have been founded already, such as the Union des Banques Arabes et Françaises (UBAF). They feel competent to invest in the oil industry; they have begun in the refining and carrying trades; but they would like to divest themselves of some of the responsibility for production; the Shah has invested in North Sea oil prospection. Many report that their ambitions to participate in Western businesses often come up against Western prejudice against letting control pass into foreign hands; they reckon that this prejudice will have to be swallowed.

The Kuwaitis here also are ahead of the other desert states, and have for years both on public and private account bought large holdings on the world's stock exchanges and shares in its trading ventures. Less experienced states have preferred short term bank deposits. The scale of the surpluses starting to accrue to them in 1974 is so new that for the present, all are doing only what they are used to doing. But all know that as their surpluses mount far beyond what they can spend, they will have to lend, and that new techniques are required. "Have you any proposals?" they ask their customers.

[1]Provisional allocation of loans to this fund was at once also made by Abu Dhabi, Canada, Kuwait, Libya, Oman, Saudi Arabia, and Venezuela.

[2]In Kuwait, where it is largest in proportion to total reserves, private assets held abroad were thought to be between $3 billion and $4 billion at the beginning of 1974. (Source: *Financial Times* (London) "Survey of Kuwait," February 28, 1974.)

IV.

Impact on the
Industrial Consumer

Fortunately for industrial consumers, they are not asked immediately to pay for all their oil imports in real terms. If they had to do so, they would be confronted with a "resource problem," meaning that no matter what their level of output and employment, they would have less resources than before for domestic use. Otherwise put, imported energy would cost their public its durables. The consequent switch from manufacturing goods for home use to manufacturing export goods would result in a drop in their standard of living, a difficult reallocation of resources to the export sector, and other painful adjustments. Thanks to limitations on the oil producers' capacity immediately to absorb imports, consumers get a reprieve from this fate. In effect, they are for the time being buying one-half or perhaps two-thirds of Middle East and North African oil on credit, and credit is a way of buying time.

Time is what they need in order to adjust to paying in real resources later. The effort required of them will be mitigated by a rise in export prices — the inevitable consequence of a sudden and significant increase in export demand — when their external terms of trade, worsened in 1974 by the oil price rises, will begin to improve from the low point then reached (unless, of course, further oil price rises are imposed by OPEC to offset the rise in the cost of its members' imports). The effort may also be mitigated because buying on credit is advantageous in an inflationary

39

world; when money rates of interest do not fully compensate for the rate of inflation, with the passage of time the debt in real terms begins to shrink. But simultaneously, the consumers will need all the time vouchsafed to them in order to make necessary adjustments in their export pattern and way of life.

The burden of the oil price rise and the adjustments that they must make to meet it is unevenly spread. Industrial consumers who themselves produce part of the oil they use, or who export some other commodity whose price has soared, can pay the new price without too much disturbance of their home economies. Canada, for instance, is a small oil importer and can do so; the United States, once it corrects its reckless consumption habits, can do the same thanks to the strength of its economy and its wealth of natural resources. Germany is managing, despite the oil crisis, to maintain a balance-of-payments surplus. At the other end of the scale Japan, which imports 86 per cent of its energy requirements and where oil accounts for three-quarters of this total, is hard hit. So are the United Kingdom, France, and Italy.

The countries worst hit feel the blow the more keenly because the oil price rise took place at a time when most were beginning to experience serious imbalance of external payments, and when the world economy was already suffering from inflation (see table 3). The pre-October 1973 balance-of-payments deficits of countries such as Italy and the United Kingdom reflect basic economic and social problems at home; countries in such precarious balance and with economic problems that tend to fester become less creditworthy in the eyes of potential lenders.

Industrial countries have therefore to start thinking out their options. These options, none of them mutually exclusive, include imposing the austerity practices already described. They also entail preparing for the day when oil producers can take more payment in export goods. The onus of bringing about these changes will fall on govern-

TABLE 3

Commodity Price Indicators: 1970 = 100

	End 1971	Oct. 1972	End 1973	Peak 1973-74
Food	90.3	136.1	242.6	280.8
Fibers	101.8	186.8	325.2	363.2
Metals	72.8	82.1	167.4	244.9

(Source: *The Economist* Sterling Commodity Price Indicators)

ments, who will have to achieve them by means of a mix of fiscal, budgetary, and monetary policies.

The predicament of the governments of the industrial world is how to attain the optimum mix while maintaining the full employment that is their major target. A government can correct the deflationary trend of successive oil price rises (which are equivalent to a tax imposed from abroad in that they reduce the level of home spending-power) by cutting domestic taxes, thereby rescuing its home consumers from discomfort and keeping up their spending power; in other words, it can subsidize its consumers so as to avoid deflation and unemployment. But in doing so it generates two unsatisfactory developments: the oil price rise, though deflationary of demand, is inflationary of the price level, and so prevents that government from reducing the rate of increase in the general price level. It also increases the deficit in its balance of international payments.

Alternatively, or simultaneously, it can meet its problem by borrowing abroad, for instance from the oil producers (as Britain, for one, has done from Iran) and use the proceeds to increase public expenditure, possibly — as France has done — developing new sources of energy in

41

this way; it can thus strive to maintain full employment, but again it increases its deficit. Either way, creditworthiness is reduced.

A suggestion made by some economists is that governments should mitigate their problem by encouraging private capital formation — that is, stimulating the corporate sector into doing the borrowing in order to increase productive capacity, contribute to exports, and lessen the nation's deficit. At first sight, the idea has attractions; it builds up a nation's capacity to pay its debts; it cures the chronic deficiency in investment activity that has been characteristic of both Britain and France, and it removes some of the responsibility from government shoulders. But second thoughts show that the relief would at best be minimal. Few industrial firms would be inclined to borrow abroad at a time of great uncertainty about fluctuating exchange rates, or to expand their plant when expecting a fall in domestic consumption.

During the few years' grace that consumer governments are going to get, their major problems are going to be of the financial kinds discussed in the next chapter. But they will also need, in addition to keeping unemployment at bay by the mix of means just described, to organize the structural changes in their industries that will enable them to pay off as much as possible of their debts in real terms — that is, in the investment goods that the oil producers should by then be ordering in quantity. People, policies, and plant cannot be caused to change overnight, but they will have to change over the years in question, and that change will not be marginal. Some retraining of labor and some curtailment of home consumption are inevitable. The weaker the political leadership, the more difficult the necessary sacrifices and adjustments will be to institute. They will be resented because they follow an unprecedented era — 25 years — of continual economic growth and improvement in living conditions. The public expects both to continue. Further, the adjustments come at a time when the militancy of many sections of the community has increased because of inflation, and when each group is attempting to protect its real income from the erosion both of actual and anticipated price rises.

In theory, the need for drastic changes should not be exaggerated. Germany, for example, has already devised means of coping with its workforce and its export trade. Others could, if prepared to face two to three years' curtailment by about one-third in the growth of consumer demand, recover their previous rate of growth and by, say, 1977, cause the deficit on balance-of-payments to start to disappear, begin to increase investment in industry, and make ready for the expected demand for investment goods. As things are, world inflation and the worldwide increase in commodity prices have made this program difficult to follow.

In practice, the changes must be made. They might be more profitable than expected because, as will be seen in Chapter VI, the oil producers seem bent on laying out sizable amounts of their surplus funds on grants and loans to poorer developing countries. Industrial countries will relieve their own dilemmas if they are ready for a gradual increase in export demand not only from oil producers but also from countries of the Arab world, the Indian subcontinent, and Africa, whom those producers have financed. The larger the amounts thus plowed into less developed countries in need of industrial and agricultural equipment, the smaller the global deficit of the industrial world. Triangular arrangements between an oil producer (for example, Saudi Arabia), a developing country (Egypt), and an oil-importing country (Japan) in which the first provides capital, the second the investment opportunity as well as labor, and the third technology and equipment may, even if hindered by obstacles, prove promising all round.

Implicit throughout this chapter is the patent truth that countries in deficit cannot maintain or even manipulate their levels of consumption and employment without borrowing, and that they must therefore contrive creditworthiness to the best of their ability (see Appendix, Table

V). From banks, Italy is finding this difficult. The United Kingdom is hoping to bridge its financial gap by borrowing extensively on the strength of future earnings from North Sea oil. (It could be caught short were the price of oil to have fallen when the loans fall due for repayment.) Meanwhile potential lenders in surplus, who are the United States and Germany, see the fight against inflation as a major feature of their economic policy, and so are shy of encouraging capital movements toward countries in deficit out of fear of the inflationary impact of simulating economic activity at home and abroad. All the same, past experience of other financial embarrassments suggests that rather than see one country collapse and start a "domino" process, rescue materializes and the first casualty gets pulled to safety out of a combination of self-interest and goodwill.[1] For the temptation to governments in an impasse is to restrict imports and engage in competitive devaluations in order to increase exports to the rest of the world — a beggar-my-neighbor policy that amounts to transferring unemployment abroad, and that can induce a world recession.

In a world of floating exchange rates, however, some stimulus to the exports of deficit countries (and some reductions in their imports) may take place without direct policy intervention. They may be induced by market changes in parity. The currencies of countries in surplus on current account will tend to appreciate relatively to others, not simply because of that surplus but because those countries will tend to attract capital inflows from oil producers. The strength of the United States dollar and the German mark will no doubt be enhanced in this way. The effect of their strength should be to stimulate exports from countries in deficit, and to damp down their imports. But these adjustments do not always take place, and when they do so, entail long time lags.

A point that is minor by comparison with those just set out is that some industrial consumers fear the competition that may develop if, with the help of oil money, industries

44

in the same fields as their own are set up in the southern Mediterranean and Asia. The development may in fact turn out to be beneficial because it will give rise to real growth in the rest of the world.

Clearly the industrial consumer, much as he would like to pay for his oil with paper claims, must in the end pay for it in real assets. His problem of the moment, however, is one of handling financial transfers, because the oil producers cannot immediately take payment in real resources. The country with by far the largest surplus funds — Saudi Arabia — may not be able to do so for years, if ever. Thus the problem of dealing with financial transfers of the oil money will be with the world for far longer than the six years that are the main concern of this report.

[1]A loan of $2 billion from Germany to Italy for a period of six months, renewable for another 18 months, was announced on August 31, 1974.

V.

Financial Problems for Producer and Consumer

The immediate problems that stem from the quadrupling of oil prices in late 1973 are in essence financial and affect three participants in the oil money game — major industrial countries, oil producers with surplus funds, and the large private international banks. The oil price rise entails financial transfers of unprecedented scale between two groups of countries; first, the outflow from consumers to producers in payment of the oil bill, and second the inflow deposited by producers who are in financial surplus on international financial markets. The financial problem is, fundamentally, the recycling of these funds from economies in balance-of-payments surplus to economies in deficit. At present, this recycling is effected by banking institutions, and these are imperfectly adapted to the dimensions of the new flows.

The financial problem is not a short term, once-and-for-all affair. The widely accepted view is that some producing countries will accumulate surplus earnings for at least ten or fifteen years (see Appendix, Table I). It is not unrelated to the transfer of real resources described in the last chapter, because oil producing countries are acquiring claims on goods and services that they intend to realize in the future. The financial problem, therefore, has implications for the economic performance of countries in balance-of-payments deficit who may prove unable to borrow the full amounts necessary to maintain their past levels or trends of economic activity. Some of these countries, because of their significance in world trade and the world's economy, may end by exporting their recession, and driving the rest of the world down the deflationary spiral.

If OPEC were a single country and all oil consumers another single country, recycling would raise no special problems. The funds received by OPEC would necessarily return to the single consumer country in amounts equivalent to its deficit. But consumer countries, as has already been pointed out, are a heterogeneous group. They differ as to dependence on oil imports, as to their balance-of-payments position, and as to their ability to secure foreign loans or attract foreign investment. It follows that the international financial system does not necessarily recycle surplus funds to oil consuming countries in accordance with their requirements.

As of mid-1974, the financial transfer mechanism operates broadly as follows:

Oil companies settle their bills to producing countries mainly in US dollars and pounds sterling. Until the recent changes in participation agreements (see Chapter I), the rough proportions between the two currencies were 75 and 25 per cent. The share of sterling may decline as a result of these agreements, for they increased the share of buy-back oil (which is paid for in US dollars) as a percentage of total off-take.

Until mid-1974, the producers tended to hold most of their surpluses in short-term deposits: exceptions were some purchases of medium- and long-term securities, small ventures into real estate, very small holdings of equities, contributions to the IMF recycling fund, and a few government-to-government loans. Saudi Arabia may be holding a larger proportion of its funds in United States government securities than do other oil producers; but as its surpluses are very considerable, the amounts that it deposits short-term are also extremely large. The bulk of OPEC surpluses as a whole are held in US dollars in New

York or in the Eurodollar market, in which London has a prominent place. A share — perhaps 20 per cent of total surpluses accumulated up to mid-1974 — may have gone into sterling; smaller amounts have been lodged in other currencies. Sterling derived some advantages from this situation (hence its relative firmness in the first half of 1974 despite the worsening of the British balance-of-payments deficit). But these advantages may not continue to accrue for very long. The major implication of the developments of end-1973 in the realm of oil money is that the world is firmly back on the dollar standard. In a sense, the US dollar has become the strongest currency. But the responsibilities of the United States *vis-à-vis* the rest of the world have also increased. Its position now in some ways resembles that it enjoyed immediately after the Second World War.

The producers' tendency to lodge their deposits short term offers them certain immediate advantages. Interest rates are high, and they value liquidity because it leaves them options for altering the disposal of their funds — to take advantage of better investment opportunities, to shoulder some important new commitment, or to parry some threat to an investment. In any case, their tendency to deposit short term is partly the result of habit. Their great wealth is very new, and they need time to adjust to it, to digest the perplexing flood of suggestions they are receiving, and to take broader decisions about a comprehensive long-term disposal of their money surpluses. They find the magnitude of these as surprising, and the implications of its size as serious, to them as they are to the rest of the world. As was explained in Chapter III, their decision-making is constrained by shortage both of expertise and of institutions — two handicaps that may delay their formulation of investment policies and other adjustments to their new situation; these delays may be long in the case of the country with the biggest surpluses — Saudi Arabia.

Oil producers are depositing their money with a handful of international banks. Some reckon that Saudi Arabia deals with 30 banks in all, but it seems that the bulk of

49

deposits from OPEC as a whole is concentrated in only some 10 or 12 institutions. Up to September 1974, the banks had coped well with the new volume of funds — well enough to induce government complacency, particularly in the United Kingdom. But the bankers themselves were worried, and had begun to doubt their ability to continue to perform to the satisfaction of oil producers, and of consumers in need of loans.

It is easily forgotten that, up to that date, only two quarters' bills had been paid, and that the ability of banks to cope with recycling the funds had therefore not been fully tested. But deposits accumulate rapidly. As this report is being written, another quarterly bill is being settled; by the time it is printed and circulated, a fourth will be due. The banks face technical problems. For one thing, they are subject to legal requirements governing capital/deposit ratios. Increases in deposits call for an increase in their equity capital, and, in the state of the capital market of September 1974, they may find it difficult to sell new stock. It is estimated, for instance, that the First National City Bank of New York, to maintain the required capital/ deposit ratio, needs $85 million per month of new capital to match the current expansion of its credit and deposits. Another technical problem is the matching of loans and deposits of differing maturities within the limits dictated by sound banking rules. The demand for loans tends to entail longer maturities than does the supply of credits.

The ability of banks to accept short-term deposits and at the same time to respond to a demand for long- and medium-term credits was, by September 1974, reckoned to be already overstretched. In principle this issue ought not to worry banks too much. Excess supplies of short-term deposits should by rights bring short-term interest rates down, making longer term investments relatively more profitable. Depositors might be expected to some extent to respond by altering the term-structure of their investments, thereby helping to improve the ability of the banks to match deposits and loans of differing maturities. But the monetary policies pursued by the United States in the third quarter of 1974 prevented a fall in interest rates for money on call. In order to fight inflation, the American Federal Reserve Bank and the United States Treasury were using an instrument — high interest rates — which had some most significant international repercussions. In the context of oil money, their policy reinforced the inclination of producers to deposit short. It reduced the scope for recycling oil money to consumer countries in balance-of-payments deficit, since their demand for medium-term loans was not matched by supplies of deposits of similar maturities.

It has been argued that deposits of oil money, though on call and open to immediate withdrawal from any bank, are in fact deposited long term if the banking system is regarded as a whole. Money withdrawn from one bank is likely to be redeposited in another. Moreover, even if some funds are withdrawn from the system, they will soon be replaced by larger amounts; for the situation is a dynamic one in which surplus earnings are continually accruing, and the net balance of deposits must inevitably grow for many years to come. These considerations point to the possibility of organizing a swap arrangement between international banks, which would cover fellow-banks against the consequences of a sudden switch of deposits. But the structure of the international banking system — which, unlike national systems, does not have a lender of last resort able both to control operations and to bail out banks when in difficulties — renders international swap agreements difficult to arrange. There is, of course, the interbank market which, informally, performs the required functions; but it affords banks no assurance that they will be able to borrow at rates that absolve them from considerable losses on some day of need. The interbank market is no sure cushion when some bank has to meet a significant withdrawal of foreign short-term deposits.

Consumers wishing to put oil surpluses deposited with them to work to the best advantage of all concerned come up against the intrinsic imperfections of the world's capital markets. Loans do not flow to those whose economies need them most, but to the creditworthy. Recycling in the interests of world prosperity calls for rules on loans and a criterion of creditworthiness different from those that have hitherto prevailed. Some argue that these rules and this criterion are in practice not very rigid, because the standard of creditworthiness is that of the marginal borrower of a given supply of loanable funds. When the supply is short, the standard of creditworthiness rises; when the supply is abundant, the criterion is automatically relaxed. Some disagree. Few banks, if any, were willing to lend to Italy in the summer of 1974. The collateralization of gold had little effect because of *pari passu* clauses on earlier loans which extended advantages secured by new lenders to earlier lenders. Italy has not enough gold to back the whole of its international debt. Its difficulty in borrowing exemplifies the recycling problem. A recession in Italy, combined with trade restrictions there, would send out ripples that might affect its main trading partners in the EEC. Germany, despite the strength of its economy, might in the end feel the impact of a contraction originating in Italy, subsequently affecting France and the United Kingdom, and aggravated by mutual repercussions on these and other economies.

Bankers are aware of these snags, but argue that they cannot step into the shoes of governments and assume responsibilities far outside their normal commercial role. They may be deriving benefit from the expansion of their activities thanks to the oil money, but this point is incidental to the argument. An increase in their profits may throw the limelight onto them, just as it was thrown onto the oil companies in mid-1974, but need not alter anyone's perception of their functions.

It is clear that on present premises and in present conditions, the recycling of funds lodged by the oil producers cannot be effected to the best advantage of the world's economy. Concepts will have to change. Some throw the ball to Central Bankers. In international monetary affairs and negotiations, Central Bankers have for some time tended to sit at the elbow of Ministers of Finance. They have not taken the lead. Further, some Central Banks do not exercise much control over the international banks located in their countries. For example, the attraction of London for many international bankers is partly explained by the policies of the Bank of England, which are very flexible in this respect. The Eurodollar market is largely unregulated by central monetary authorities, which explains both its past expansion and many worries about its behavior. Yet a handful of Central Banks — for instance, those of the United States, the United Kingdom, France, and Germany, acting in concert — might succeed in enhancing the ability of private international banks to cope with the recycling problem. They would be in a position to arrange swap agreements, after imposing a minimum of necessary controls, and to define a lending policy with the guarantees that private banks might seek. But, save for a statement by the German Chancellor, there are at the time of writing no signs of such a move.

Others want to throw the ball into the oil producers' court, arguing that it is up to them to recycle money in the best interest of the world's economy. But this principle is unacceptable to them. Creditworthiness matters to them when they lend, because their surpluses are assets obtained in exchange for a depletable resource — their oil. If they fail to invest these assets safely and profitably, they fail in their obligations to future generations in their own countries, and expose these to the risk of a return to poverty when the oil era is over. They are ready to help developing countries, but reckon that responsibility for the well-being of the industrial world falls on the industrial countries themselves, and particularly on the United States and Germany — the strongest and least adversely affected among them.

Yet another suggestion is that the governments of the consuming countries should market their debts internationally. The international banking system could in this event retain a function, but mainly as a jobber. In any case, international banks face problems in lending to governments. Lending to those with small margins may increase their total profits, but lower their returns on equity. Loans with higher margins are not attractive to governments. And banks do not like their balance sheets clogged with long-term loans.

Domestically, governments do not in the main run their national debt by borrowing from banks. Logically, the same principle should apply in the international field. In the new era in which they have increasingly to seek long-term credit abroad, they should be selling debt instruments to other governments (which are the international counterpart of their home public). This practice calls for a change of attitude about management of balance-of-payments deficits. These would have to be regarded not, as at present, as a cyclical or annual problem, but as a secular one. Conventionally, secular debt is thought of as unmanageable. Yet there is no way of avoiding the prospect of secular debt, and the conventional view about its management will have to be revised.

At the Rome seminar, strong arguments were advanced in this context in favor of multilateral lending between governments; hence the role reserved to international banks as intermediaries. Some were against bilateral government-to-government deals, reckoning that they had the marked disadvantages that are set out in Chapter VII. The views expressed were by no means unanimous on these points. All were agreed, by contrast, that adjustment to the new situation entailed an act of political will, and that, to date, too few governments had thought in terms of the adjustments that they were bound to have to make. Some were passive about them, others complacent; yet others had not yet realized the dimensions of the financial problem with which both they and the international banks were confronted.

VI.

The Small Consumers: Bailing Out The Less Developed Countries

The theory that industrial consumers cannot pay for oil at the new price without wrecking their economies becomes hard fact when applied to the less developed countries (LDCs). Industrial consumers in Europe and North America have some latitude for curbing the home consumer, or shifting investment from sector to sector in order to meet their oil bill. Most LDCs have no such flexibility. The bulk of their people live on or below the breadline, and to ask them to change their work or even their diet may well condemn them to starve. These countries are hit by the rise in price not merely of crude oil and the oil products that include fertilizers, but by the effect of the oil price on the import cost of the machinery that they need if their industries — agriculture in particular — are to grow. Many of them are beset by a very high rate of population increase. Already, they are forced by the price of oil to curtail their overall growth; to suggest that they should do more in this line is to underrate the revolutionary forces that their poverty could set loose in their area. To give figures: in 1973 their consumption of oil and products cost them $5 billion; the estimated figure for 1974 is $15 billion. The difference between the two figures comes to more than the total aid from official sources — $8 billion — that they received in 1973.

A few LDCs escape these hardships. Some produce oil locally; Brazil, Malaysia, Argentina, Brunei, and one or two others cumulatively produce over one million barrels a day, and satisfy their own needs or some of them. Others are major producers of raw materials whose prices have substantially increased in the last few years, such as cocoa, sisal, sugar, rubber, copper, or phosphates.[1] These have reserves on which they can draw in order to pay for oil; also, they are reckoned to be good investment risks, and so easily raise loans and development aid; "to him that hath shall be given." But from mid-1974 the price of most of their products started to fall, both absolutely and in relation to the price of manufactured goods, and to that of oil. As soon as their reserves are exhausted, they will have as little room for maneuver as the rest of the LDCs.

A description of India's fortunes will serve to illustrate the plight of the worst afflicted. India is not a major importer of oil; it consumed 250,000 barrels per day of crude in 1973, and the equivalent of 46,000 barrels per day in products. It produces in Assam and Gujerat about one-third of its total oil consumption; though it exports primary products, the price of none of these has risen on the dramatic scale that would help to pay its oil import bill. Its nonessential use of oil is negligible. When it set its targets for its Fifth Five Year Plan, (the fourth was to end in March 1974) it expected 50 per cent growth over the new quinquennium, and anticipated a growth in oil consumption from the 1972 rate of 460,000 barrels per day to 720,000 by the end of the decade. Then came the quadrupling of its oil bill; whereas oil constituted 11 per cent of its total import bill in 1972, the figure for 1974 became 45 per cent of the total, and at $9 per barrel would become 55 per cent by 1978. To pay such a bill in 1974 without worsening its balance-of-payments position, an immediate rise of 30 per cent in its exports would be necessary. The figure is clearly impossible of achievement.

India at once took steps to meet the crisis. It raised the price of gasoline by 75-80 per cent, so bringing about a 25 per cent reduction in consumption. By exercising its import licensing system and the allocation of foreign exchange, it held the total consumption of oil and products down to the 1972-73 level. Its main uses of oil are for transport, industry, cooking, lighting, and fertilizers (it gets nearly all its electrical power from coal) and its targets are a quick reduction of consumption by 40,000 barrels per day of crude and 20,000 of products. Even with this degree of self-discipline, it faces the hardships of a lowered standard of living, a lowered rate of growth, and cuts in food production. These prospects are so disastrous that not only India but all LDCs in like distress — they include Pakistan, Srilanka, Bangladesh, and several Central African states — must at once find ways of alleviating the social and political trials entailed. Means that struck them forthwith were direct arrangements with friendly oil producing governments, resort to international organizations for more aid, and cheap loans from any quarter.

Some of the means that occurred to them have been discarded or had best be so. One was differential price levels for oil. This expedient is no longer thought possible, largely because the producer states, who would have to set the scale to be charged to their various customers, do not want the burden of drawing invidious distinctions; also, the producers saw in the granting of special rates to friends a risk of creating rifts within OPEC; lastly, the plan opened up prospects of black market operations if customers granted oil on the cheap chose to sell it instead of using it. Another suggestion was loans granted piecemeal on concessionary terms to alleviate particular balance-of-payments difficulties. The drawback to these is that they are not directly linked to increasing productivity and creating assets with which to pay the interest and fund the loans.

Useful means of succor include:

1. Bilateral program or project loans from the producers. (e.g., KFED loans at ½ to 4 per cent over 15 to 50

years, or Iran's $1.2 billion loan to Egypt for 25 years at 2 per cent for rebuilding in the Suez Canal Zone area.)

2. Bilateral loans linked to increasing the borrower's export potential as well as increasing his capital assets. An example is the Iranian-Indian oil deal that includes a loan to India of $300 million for the import of equipment for mining and pelletizing India's iron ore, and for producing bauxite and alumina. Part payment for the oil is to be made in installments over five years at concessional rates of interest, part through the export to Iran of pelletized ore and alumina within five years of the completion of the sponsored projects.

3. Loans from a Regional Assistance Fund for Economic and Social Development of the kind set up by a group of the oil producers in 1971 for use in the Arab world, with its seat in Kuwait.[2] An advantage of such funds is that by subsidizing poorer neighbors they promote purchase of goods manufactured in the West and Japan, and by means of triangular trade help to promote payment of the oil bill in ways attractive to the producer.

4. Loans from international institutions. Only the richer LDCs can afford IBRD's (August 1974) lending rate of 8 per cent. Its total loans, including those supplied through its International Development Aid (IDA) total only $4 billion — a small sum when set against the $9 billion of additional oil costs to the LDCs in 1974. One suggestion is that IBRD bonds would become a better tool of development assistance if the Bank were to lend at a lower rate than that at which it can borrow from the oil producers and other rich clients, and if the industrial states of the Western world were to agree to make up the difference.[3]

5. Agreement with the producers on deferred payment terms. (A past example is the Iranian-Indian oil-deal described in paragraph 2.) These schemes are more

popular with the borrower than with the lender. They offer the former a chance of paying for his oil when less crippled by inflation than he is at present. Like all bilateral loans between governments, they expose the lender to requests for renegotiation, to thinly disguised reneging, or to hazards connected with political change during the life of the agreement.

All LDCs know that they cannot raise their growth rate without soft loans; all are hoping for these not only from the oil producers but also from the industrial consumers with whom large amounts of surplus oil funds have been lodged; such triangular deals could benefit everyone concerned. Some LDCs are hesitant about loans from individual states because they fear that (as in the past with imperial patrons) the lenders may wish to attach strings to their bounty; others are mistrustful of international sources because so many of these are under Western management, and most are so slow to get into gear. Nevertheless they must resort to one or other; their problem is to appear creditworthy.

The oil-producing states, as has been shown, are full of good-will towards the LDCs, and intend to invest there (see Chapter III). Those ready to take action at once are urging LDCs to set and proclaim their short-term priorities now. They know that investment in some LDCs constitutes a risk, but it is a risk that Kuwait and Iran in particular have shown themselves ready to take, as did General Marshall's United States, in order to buy themselves an insurance against social and political havoc in their area. The essential, in the interests of both parties, is that the LDCs' export potential should rise so that they can pay for oil in goods, and that the producers, by investing there, should buy themselves some real assets in Asia and Africa.

The volume of development assistance to the LDCs that some producer states say that they are contemplating is enough to make the Western world blush. For the whole

59

period since the Second World War, its members have been the major source of aid to what they then called the under-developed world. Their feats, performed both in their individual political and economic interests, and sometimes apolitically through international funds and charitable institutions, do them credit, but have never been up to target. Their contributions have never, for instance, reached the level prescribed by UNCTAD, which is 1 per cent of their national income.

Yet they are on common ground with the producer states if they continue to bail out the LDCs. Large new contributions by the latter will not exonerate the industrial states from contributing because, riches for riches, they are still the better off of the two. What is more, producer contributions to poor countries will benefit industrial countries also, since peoples in need will be better able to buy their manufactures.

There is as much ground for combined operations as for competition in all less developed areas. In sum, there are limitless chances of marrying the oil producers' wealth to the Western world's technical ability.

[1]e.g., Multiplication factors. End 1971 to peak price 1973-74: Cocoa, x 7; Sisal, x 6; Sugar, x 5; Rubber, x 4; Copper (wire bars), x 3; Phosphates, which are, like oil, held in few hands, tripled in price from one day to the next. (source; *The Economist* Sterling Commodity Price Indicators).
[2]Not fully in operation in mid-1974 as the Saudi contribution had not been received.
[3]For IBRD's temporary oil emergency fund see above, Chapter III p. 36 and for special interest rates, Chapter VII p. 65.

VII.

Outlets for Surplus Funds

In the years to 1980 that were the Rome seminar's main concern, the shift of the world's liquidity to the Middle East may be at its zenith because so much of the money paid to the oil producers cannot immediately be absorbed at home.

As the avalanche of new money swells to dimensions that the world's banking system cannot handle, both producers and consumers will, if they want to avoid the fate of the sorcerer's apprentice, have to seek a wide range of outlets for the surplus funds. The importance of doing this is paramount. Success in doing so will not only help the producers to maximize their asset. It will insure their willingness to supply the world with oil. It will also save both producer and consumer from the risks of economic instability and galloping inflation that could stem from an expansion of world liquidity on a scale never as yet experienced.

The Rome Conference made an attempt to draw up a list of outlets. It left aside spending on OPEC's domestic business — on a better infrastructure, defense and internal security, social improvements, wider income distribution, job creation, and the establishment of new home industries such as refineries, tanker fleets, airlines, shipping lines, and the like. It also left aside the attraction of foreign industries into OPEC states — a process already in full swing in several countries. It confined itself to outlets abroad, most of them assets offering the three attributes that producers want most — security, a real rate of return, and marketability.

Outlets immediately available to most OPEC producers were listed as:

1. Payment of foreign debt — a big item for Algeria, Indonesia, and Nigeria.

2. Regional development, amounting to the purchase of real assets in the producer's own region. The close involvement of the producers in financing development in the region was described in Chapter III.

3. Purchase of real property abroad, long practiced by Arab producers in Cairo and Beirut and now being extended beyond the Arab world.

4. Purchases of equities and equity assets abroad. Kuwait maintains an investment office in London for making both initial and long-term placements, and owns portfolios managed by major banks in the United States, Germany, and Switzerland. Representatives of the producer states notice resistance by nationalist consumers to the purchase of large holdings of equity assets. Also, some producer states have not had time to get used to the risks of moving into equities; the higher yield does not offset their dislike of making investments over which they have no control.

5. Entry into new joint ventures with Western and Japanese industries and banking houses.

6. Bilateral government-to-government loans. These are both fashionable and popular with the producers. But the system has some drawbacks:

    The borrowing government externalizes its national debt, and sees this traded in by a handful of foreign governments.

The greater creditworthiness of the more efficiently run consumer states will mean that they, and not the states in economic difficulties, will get the best terms.

Intergovernment debt is politically sensitive; past experience suggests that reneging is common.

There are complications about the currency in which such debts are arranged; if in that of one of the desert producers, he can revalue with little or no effect on his own economy.

7. Development assistance and soft loans to less developed countries. Producers who lend or invest thus do so at risk, but it is a risk that some wish to take for psychological or political reasons, preferring it to investment in industrial countries for reasons described in Chapter III.

8. Rebate and deferred payment arrangements to help LDCs. Some producers are shy of these because of the risk that recipients come to see them as gifts.

9. Development assistance and credits applied to LDCs through international bodies; IDA, IBRD, and national institutions internationally known and accepted (the Export-Import Bank, ECGD, etc.). The Shah of Iran lately proposed the creating of a new "neutral international organization" for this purpose, to be run by OPEC, consumer and recipient representatives in equal numbers, and to handle a fund of $3 billion annually put up in equal shares by 12 OPEC members and 12 industrial countries. Some OPEC donors prefer to deal individually and directly with LDCs as being a quicker process.

At the Rome meeting, some new devices, as yet untried in the special field of the oil surpluses, were suggested. None were intended to apply exclusively to oil money, since many producers object to investing or lending under arrangements made for them alone.

1. Indexed bonds. By indexing in a way designed to offset all risk of devaluation or inflation, such bonds could become meaningful paper claims, with tomorrow's real value related to that of today.

   Possible indexes include a basket of imports that the consumer wants (a mix of capital goods, consumer goods, food, etc.). A drawback is the well-known technical difficulty of constructing acceptable price indices.

   No index is ideal. Another possibility is a floating rate bond with a variable rate of return; the United States banking system may already be moving toward this, with bonds redeemable immediately and yielding interest pegged to 1 per cent over the three month Treasury Bill rate.

2. A world Unit Trust or Mutual Fund, or possibly a series of such trusts, with a form of direction that would have the confidence of the contributors.[1] These would be international institutions set up by treaty to buy equities and property across the world. No one holder would hold more than a small proportion of the total, and agreement would be needed about the pace at which he could redeem his share; multilateral guarantees against confiscation would also be necessary. Such funds would be open to governments other than those of the oil-producing states.

   The advantages of the scheme are that:
   it reduces the liquidity problem;
   it encourages new investment;
   it does not give the foreign investor direct control in institutions sacred to the industrial world;
   it gives oil producers a stake in the growth of the world economy;
   it links producer assets to world performance.

The drawbacks are that:
    it is legally cumbersome to set up;
    it entails complex negotiations to enable it to handle the tax laws, disclosure requirements, etc. of the states in which investments are made;
    it entails an enormous cash flow that might disturb and raise the price of equities.

The relevant machinery would probably not be worth inventing if the need for such an institution were temporary. As things are shaping, the need will be permanent, and the device seems worth trial if it can be started to meet the needs of the 1970s.

3. World Bank or other international bonds borrowed from the producers at a commercial rate of interest and lent to LDCs on concessionary terms. The difference between the two rates of interest to be met out of an Interest Equalization Fund to be supplied by the industrial consumers, on whom the burden would be small in proportion to the cost of the aid they have supplied to the LDCs in the past.

4. Acceptance of payment in gold. The supply of this is not great enough to make a big dent in the surpluses, but stocks happen to lie in the hands of the principal consumers. The method of payment would suit most of Western Europe. No Western holder would wish to surrender more than a proportion of his holding, and in any case, the producers would not wish to become sole holders.[2] Indeed, some might prefer not to acquire gold at all, seeing it as a more speculative commodity than oil. To reduce the risk entailed, definition of price and/or remonetization would be necessary.

No list is exhaustive. No one of the outlets listed above will, on its own, solve anyone's whole problem. The mix adopted, and the other possibilities that may come to mind will — given the major differences between the need of the various producers and consumers — differ from state to state. But lists are well worth making because an entirely new situation calls for rethinking and invention on the part of all. Without this progress toward solutions mutually discussed and adopted, the world's economic and financial systems could grind to a halt.

[1]See I.M.D. Little and R. Mabro, "Coping with the Arab Billions," *Financial Times* (London), December 27, 1973.
[2]See F. Hirsch and P. Oppenheimer, "Selling Gold to the Arabs," *Financial Times* (London), January 15, 1974.

VIII.

Some Conclusions

For a while, the shock of the oil price rise caused consumers to dwell only on its short-term implications. Implicit in this synthesis of the discussions held in Rome are some longer term points that call for re-emphasis:

1. Throughout the final quarter of our century, two powers (excluding the U.S.S.R. and China) will dominate the oil scene. These two are the United States and Saudi Arabia.

    The United States will be the dominant consumer because its historic economic strength is unimpaired, because it is in a position to organize an independent energy policy, and because of the seeming inability of Europe to pool its own economic strength. If Europe, beset by weak governments and mounting inflation, continues to run away from mutual solutions, the United States is strong enough to go it alone.

    Saudi Arabia dominates the scene at the producer end because its proven reserves are so much greater than anyone else's (see table 4), and because its resultant financial surpluses are so large; moreover, the expectations of its people can be far more than met by its earnings. It can therefore produce or conserve at will, and, at a time when its peers such as Iran will

TABLE 4

	Oil Reserves* (000,000 barrels)	Gas Reserves (000,000 cu. ft.)	Production** (1973: 000b/d)
Saudi Arabia	132,000	50,000	7,345
Kuwait	64,000	32,500	2,755
Iran	60,000	270,000	5,895
Iraq	31,500	22,000	1,980
Libya	25,500	27,000	2,180
Abu Dhabi	21,500	12,500	1,305

*(Source: Oil and Gas Journal)
**(Source: BP Statistical Review, 1973.)

---

be spending their whole earnings on absorbable real resources, can dominate supply and price if its regime so wishes.

2. The financial problem, which covers both the dissatisfaction of the producers with the way in which their surplus funds are handled, and the uncertainty of consumers in credit about recycling to countries in chronic deficit, is not only immediate but long term, since some producers — notably Saudi Arabia — may have surpluses to invest into the twenty-first century. World recession aggravates these worries; a world boom may not cure them.

This problem is one of institutions and political will, and banks have an interest in its solution because, without this help, they are overstretched. The fragile system of 1974 exposes them to the risk of crashes that may engulf some of them, and harm both their creditors and their debtors.

3. The problem of payment for oil in real resources could in a not so long run prove the more intractable of the two because of risk of scarcities. This can be mitigated only if governments in the industrial countries face up to preparing their peoples for some cut in their standard of living and for some reshaping of industry.

4. Meanwhile, borrowing by consumer governments in deficit is inevitable. Preserving creditworthiness is part of their task, and is of interest also to potential lenders within OPEC. Failure to recycle on the part of OPEC members and of creditworthy consumer states could induce the vicious circle of a recession.

5. OPEC members, if denied satisfaction about secure investment, or the availability of suitable goods, may well bow to domestic pressure to conserve all oil other than that sold to satisfy immediate home needs.

6. On the other hand, if offered a satisfactory package of investments for their surplus funds, including real assets, OPEC members may become willing to negotiate lower oil prices.

These facts point to areas of common interest lying behind the conflicts that divide oil producer and oil consumer. The prosperity of the former hangs in the last resort on the prosperity of the latter. Unless the world's economy continues to grow in the dimensions that prevailed before the oil price rose, the producers' gain will be less than the consumers' loss, and the result a negative or minus sum.

Both have interest in avoiding this disaster; both wish to parry a world recession; both have a direct interest in helping the less developed countries; both gain advantage if they assist countries in straits to organize self-help with balance-of-payments crises; both have an interest in insuring that surplus earnings are well enough invested to cause the oil to continue to flow.

Vital to these ends is the nature of the dialogue between them. It may not always run smoothly. It may be complicated by political factors such as United States involvement in the Palestine dispute, or tensions within the Western world, or even within OPEC. Among the paths to serene dialogue can be numbered: (1) strengthened OPEC representation on international bodies; (2) organization of official and private study groups and seminars; and (3) the holding of well-prepared study groups and conferences, of the kind already envisaged between the Arab states and the EEC, to be held now in the East and now in the West.

Appendix I.

Select Bibliography

Books

Connery, Robert H. and Gilmour, Robert S. (eds.). *The National Energy Problem*, Academy of Political Science, New York 1974.

Energy Policy Project, The Ford Foundation. *A Time To Choose*, (S. David Freeman, Project Director), Ballinger Publishing Company, 17 Dunster Street, Cambridge, Massachusetts 1974.

Erickson, E.W. and Waverman, L. (ed.). *The Energy Question: an International Failure of Policy*, University of Toronto Press, 1974.

Hill, P. and Vielvoye, R. *Energy in Crisis, A Guide to World Oil Supply and Demand and Alternative Resources*, Robert Yeatman, London 1974.

Mancke, Richard. *The Failure of U.S. Energy Policy*, Columbia University Press, New York 1974.

Stephens, R. *The Arabs' New Frontier*, Temple Smith, London 1973.

Handbooks and Articles

The Middle East and North Africa: Survey and Reference Book. 1973-4. Chapter on Oil and OPEC., pp. 63-81. By Michael Field, Europa Publications, London 1974.

London and Cambridge Economic Bulletin. "The World Commodity Boom and its Implications," Department of Applied Economics, Cambridge, England, 1974.

B.P. Statistical Review of the World Oil Industry. (Annual) 1973, British Petroleum, London.

O.E.C.D. Long Term Energy Assessment: Demand and Supply Projections up to 1985, Paris, April 1974 (Ref. ELT/CU/0240).

The Banker, London. "Oil and Money," Supplement, March 1974.

Honavar, R.M. "India and the Oil Problem," *The World Today*, Chatham House, London, July 1974.

Hunter, Robert E. "The Energy 'Crisis' and U.S. Foreign Policy," *Headline Series No. 216, Foreign Policy Association*, June 1973.

Issawi, Charles. "Oil, The Middle East and The World," *The Washington Papers 4*, The Center for Strategic and International Studies, Georgetown University, Washington, D.C., Library Press, New York 1972.

Levy, Walter. "Implications of Exploding World Oil Costs," 93rd Congress, 2nd Session, Committee on Interior and Insular Affairs (Series # 93-36(92-71) pursuant to S.RES.45, a National Fuel and Energy Policy Study).

"World Oil Cooperation or International Chaos," *Foreign Affairs*, New York, July 1974.

Mabro, R. and Monroe, E. "Arab Wealth from Oil: Problems of its Investment," *International Affairs*, Chatham House, London, January 1974

Penrose, Edith. "Origins and Development of the International Oil Crisis," *Millenium*, London School of Economics, Vol. III, No. 1, 1974.

Pollack, Gerald A. "The Economic Consequences of the Energy Crisis," *Foreign Affairs*, New York, April 1974.

Rocks, Lawrence, and Runyon, Richard P. "Energy Crisis — Society Convulsion," *Chemical Engineering News,* March 26, 1973.

Tunlir, J. "Oil Payments and Oil Debt in the World Economy," *Lloyds Bank Review,* London, No. 113, July 1974.

Yamani, Sheikh Ahmed Zaki. "Oil: Towards a New Producer-Consumer Relationship," *The World Today,* Chatham House, London, November 1974.

Zombanakis, M. "New Sources of Funds and Their Effect on the International Market." Prepared for a seminar on *New York as a World Financial Centre,* June 1974. (unpublished copies from First Boston, Europe, Ltd., 16 Finsbury Circus, London E.C.2.)

Related articles published by the American Universities Field Staff in its Fieldstaff Reports series

Franda, Marcus. *India: An Unprecedented National Crisis.* [MFF-4-'73].

   India and the Energy Crunch. [MFF-1-'74].

Gallagher, Charles F. *Spain, Development, and the Energy Crisis* [CFG-5-'73].

   The Finite Fiftieth State [CFG-4-'73].

Gall, Norman. *Oil and Democracy in Venezuela* [NG-1,2-'73].

Hanna, Willard A. *Petroleum as Panacea* [WAH-7-'71].

McLin, Jon. *Oil, Money, and the Common Market* [JM-2-'74].

   Resources and Authority in the North-East Atlantic Parts I, II, III, and IV [JM-4,5,6,7-'73].

Waterbury, John. *Land, Man, and Development in Algeria,* Parts I, II, and III [JW-1,2,3-'73].

Appendix II.

Supplementary Tables And Diagrams

TABLE I
ESTIMATES OF THE PRODUCERS'
SURPLUS FUNDS, 1974-1983
Cumulative, at 1974 constant prices ($ billions)

Saudi Arabia	280
Iraq	85
Abu Dhabi	80
Kuwait	70
Libya	70
Iran	30
Qatar	10
Total	625

These figures, necessarily guesses, include estimated home expenditure on armaments, the oil and petrochemical industries, and regional aid, though not other imports at the 1974 rate. On their basis, the amount available for investment in the world outside the Middle East over the ten years in question would be $450 billion at 1974 constant prices.

TABLE II
INTERNATIONAL OIL TRADE: SUPPLY FORECASTS
(million barrels per day)

Primary Exporters	1973 (based on Jan.-Sept.)	1980 (est.) "Normal" (i.e., no startling new developments)	1980 (est.) "Restricted" (for reasons given below)
Venezuela	3.2	2.0	1.5 Low reserves to production ratio; conservation
Algeria	1.0	0.9	0.9 Assuming no further finds
Libya	2.2	2.1	1.3 Absorptive capacity difficult to judge
Nigeria	1.9	1.6	1.6 Possible technical maximum; conservation.
Iran	5.4	6.3	3.1 Conservation
Iraq	1.6	3.3	1.3 Restricted spending or more capacity
Kuwait	2.7	2.3	0.8 Absorptive capacity; conservation.
Abu Dhabi	1.2	2.4	1.4 Rate arbitrary.
Indonesia	1.1	2.0	1.7 Low reserves to production ratio; conservation.
Other areas	2.7	5.6	5.0
Total	23.0	28.5	18.6
Saudi Arabia	7.3	10.0	5.0 Arbitrary
Total Internationally Traded Oil	30.3	38.5	23.6

These estimates were prepared in June 1974; since then, price and austerity have contributed to reduced demand and the "normal" figure is likely to be lower (see Table IV).

Exporters' home demand met from indigenous production is assumed to rise from 5 million b/d in 1973 to 12 million b/d in 1980.

TABLE III
INTERNATIONAL OIL TRADE: DEMAND FORECASTS
(million barrels per day)[a]

Import Regions	1973 (based on Jan.-Sept.)	1980 "Normal"	1980 "Restricted supply" [e]
United States	6.1	10.5[c]	5-6
Japan	5.4	7.5	5-6
Other non-Communist	3.9	4.3[d]	3-4[f]
Western Europe	14.9[b]	16.2	10.6-7.6
Total	30.3	38.5	23.6

a) These estimates were prepared in June 1974; since then price and austerity have curtailed demand and the "normal" figure is likely to be lower (see Table IV).
b) Does not include imports of 0.8 b/d from Communist areas.
c) Assumes U.S. indigenous production of 10.5 million b/d including Alaska at 2 million b/d.
d) Excludes local production in Brazil, Argentina, Malaysia, Brunei,etc., cumulatively one million b/d in 1973.
e) Assumes crisis austerity programs and fundamental changes of habit.
f) Restriction low because oil is a political and humanitarian necessity in less developed countries.

TABLE IV.
CONTRACTION OF DEMAND, 1974

	Oil Consumption Jan.-June 1974 (million b/d)	Percentage change over Jan.-June 1973
U.S.A.	16.6	− 5.2%
Japan	4.9	+ 1.1%
United Kingdom	1.9	− 8.2%
France [a]	2.17	− 5.6%
W. Germany	2.33	− 15.0%
Italy	1.79	− 2.7%

a) Heavily biased by rush in June to stock up before rationing.

75

TABLE V **CREDITWORTHINESS: WESTERN EUROPE, 1974**

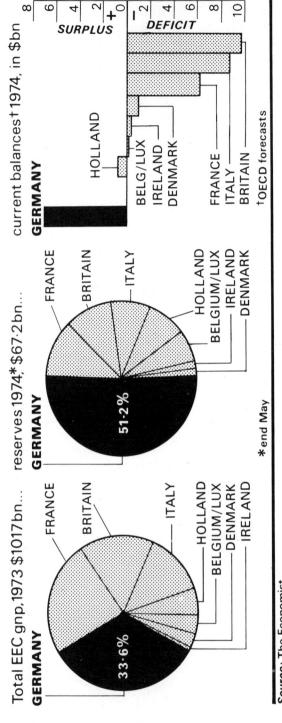

Total EEC gnp.1973 $1017bn....

reserves 1974*, $67·2bn....

current balances†1974, in $bn

GERMANY

FRANCE
BRITAIN
ITALY
HOLLAND
BELGIUM/LUX
DENMARK
IRELAND

GERMANY

FRANCE
BRITAIN
ITALY
HOLLAND
BELGIUM/LUX
IRELAND
DENMARK

GERMANY

SURPLUS
DEFICIT

HOLLAND
BELG/LUX
IRELAND
DENMARK

FRANCE
ITALY
BRITAIN

†OECD forecasts

33·6%

51·2%

*end May

Source: The Economist

76